D1161099

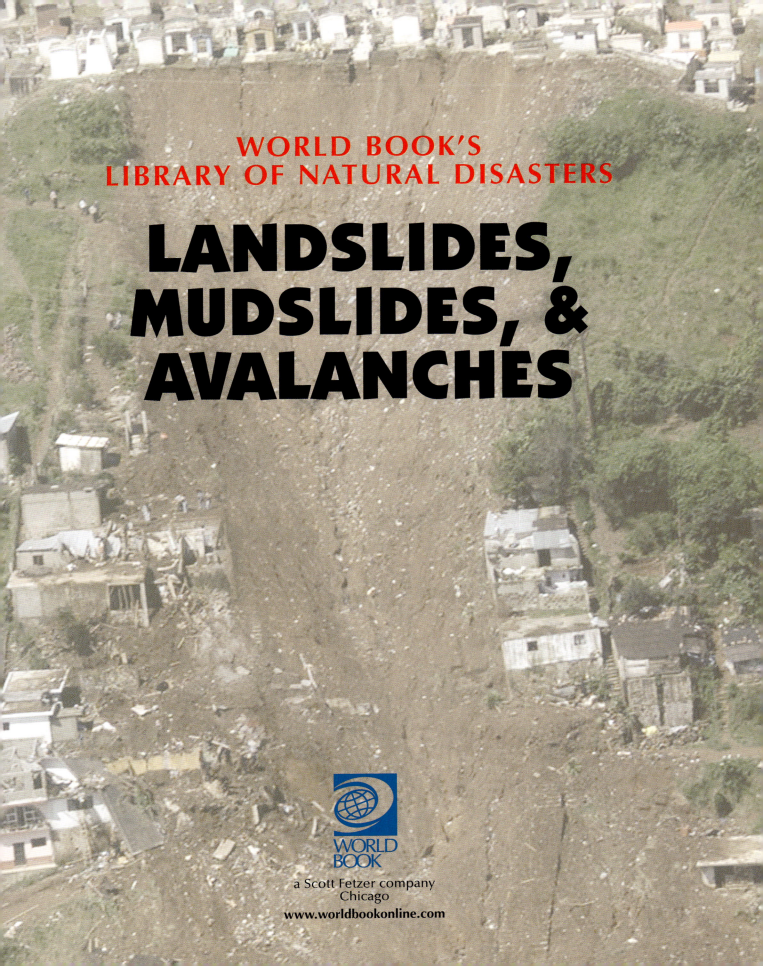

WORLD BOOK'S
LIBRARY OF NATURAL DISASTERS

LANDSLIDES, MUDSLIDES, & AVALANCHES

WORLD
BOOK

a Scott Fetzer company
Chicago
www.worldbookonline.com

World Book, Inc.
233 N. Michigan Avenue
Chicago, IL 60601
U.S.A.

For information about other World Book publications, visit our Web site at **http://www.worldbookonline.com** or call **1-800-WORLDBK (967-5325)**.

For information about sales to schools and libraries, call **1-800-975-3250 (United States); 1-800-837-5365 (Canada)**.

2nd edition

The Library of Congress has cataloged an earlier edition of this title as follows:

Landslides, mudslides, & avalanches.
    p. cm. -- (World Book's library of natural disasters)
    Summary: "A discussion of major types of natural disasters, including descriptions of some of the most destructive; explanations of these phenomena, what causes them, and where they occur; and information about how to prepare for and survive these forces of nature. Features include an activity, glossary, list of resources, and index"--Provided by publisher.
    Includes bibliographical references and index.
    ISBN 978-0-7166-9810-4
    1. Landslides. 2. Mudslides. 3. Avalanches.
I. Title: Landslides, mudslides, & avalanches.
II. World Book, Inc.
QE599.A2L365 2007
363.34'9--dc22
                            2007013813

This edition:
ISBN: 978-0-7166-9826-5 (Landslides, Mudslides, & Avalanches)
ISBN: 978-0-7166-9817-3 (set)

Printed in China
1 2 3 4 5 12 11 10 09 08

Editor in Chief: Paul A. Kobasa

Supplementary Publications

   Associate Director: Scott Thomas
   Managing Editor: Barbara A. Mayes

Editors: Jeff De La Rosa, Nicholas Kilzer, Christine Sullivan, Kristina A. Vaicikonis

Researchers: Cheryl Graham, Jacqueline Jasek

Manager, Contracts & Compliance (Rights & Permissions): Loranne K. Shields

Graphics and Design

   Associate Director: Sandra M. Dyrlund
   Associate Manager, Design: Brenda B. Tropinski
   Associate Manager, Photography: Tom Evans
   Designer: Matt Carrington

Production

   Director, Manufacturing and Pre-Press: Carma Fazio
   Manager, Manufacturing: Steven Hueppchen
   Manager, Production/Technology: Anne Fritzinger
   Proofreader: Emilie Schrage

Product development: Arcturus Publishing Limited

Writer: Philip Steele
Editors: Nicola Barber, Alex Woolf
Designer: Jane Hawkins
Illustrator: Stefan Chabluk

Acknowledgments:

Corbis: cover/ title page, 17, 24 (Reuters), 4 (Daniel A. Anderson/ Orange County Register), 5 (Prensa Nicaragua/ Corbis Sygma), 7 (Brendan McDermid/ epa), 9 (Mian Khursheed/ Reuters), 11 (Keren Su), 12 (Jebb Harris/ Orange County Register), 13 (Philippine Air Force/ epa), 14 (Yannis Behhrakis/ Reuters), 15 (Alessandro Della Valle/ epa), 18, 19 (Bettmann), 20 (Alison Wright), 21 (Hulton-Deutsch Collection), 25 (Roger Ressmeyer), 26 (Jacques Langevin/ Corbis Sygma), 29 (Robert Galbraith/ Reuters), 30 (Christophe Boisvieux), 31 (Henrik Trygg), 32 (Kit Kittle), 33 (Lee Cohen), 34 (Picimpact), 35 (Nick Hawkes/ Ecoscene), 36 (epa), 37 (Jeff Vanuga), 38 (Fabrice Coffrini/ epa), 42 (Lee Cohen).

NOAA: 16 (R.L. Schuster/ U.S. Geological Survey).

Science Photo Library: 22 (F.S. Westmorland), 23 (Lynette Cook), 39 (Dr. Jurg Alean), 43 (BSIP, Platriez).

University of Washington Libraries, Special Collections: 40 (A. Curtis 17461), 41 (A. Curtis 17463).

U.S. Geological Survey: 27 (Cascades Volcano Observatory/ Tom Casadevall), 28.

TABLE OF CONTENTS

Glossary There is a glossary of terms on pages 45-46. Terms defined in the glossary are in type **that looks like this** on their first appearance on any spread (two facing pages).

Additional resources Books for further reading and recommended Web sites are listed on page 47. Because of the nature of the Internet, some Web site addresses may have changed since publication. The publisher has no responsibility for any such changes or for the content of cited sources.

MASS MOVEMENT

Landslides, mudslides, and avalanches (or snow slides) share certain characteristics. In all of these events, masses of material slide down a slope. In a landslide, the material in question is rock, earth, or **debris.** The word *mudslide* has a less precise meaning, but it often is used to mean a landslide that is made up of material with high water content. Finally, an avalanche is a mass of moving snow.

As the mass of material travels down a slope in these events, it can pick up such items as boulders (rocks at least 10 inches [254 millimeters] in diameter); uprooted trees; bricks; trucks and cars; cement from roads and buildings; and even animals and people. The speed at which the material slides can vary. Some **debris flows** can travel at speeds of 200 miles (320 kilometers)

Houses that once stood atop the hill lie shattered below in the wake of a 2006 landslide in Laguna Beach, California. A resident recalled hearing a sound "like a bomb" as the slide began at about 6:45 a.m. While a total of 28 houses were wrecked, no one was seriously injured in the slide.

per hour, but 30 to 50 miles (50 to 80 kilometers) per hour is more common. The most swift, dramatic, and dangerous movement of material usually occurs on the slopes of mountains or steep-sided hills.

Slopes and gravity

There are many reasons why material becomes unstable and a slide begins. A volcanic **eruption** or an **earthquake** can start a landslide of rock and soil; melting snow or heavy rain can set off a flow of mud; heavy winds or explosions can begin an avalanche of snow. **Gravity** is the powerful force that pulls unstable material down a slope. The wide range of triggers that can begin a slide means that they are often difficult to predict.

Path of destruction

Landslides, mudslides, and avalanches can be extremely destructive. The moving mass of material often destroys everything in its path, smashing buildings, burying houses, and injuring or killing people. The damage caused by landslides worldwide costs billions of dollars each year. The damage caused by landslides in the United States alone is, on average, around $1 billion to $2 billion per year. In addition to the monetary cost, around 25 people are killed annually in the United States by landslides.

A girl is airlifted to safety from a mudslide in Nicaragua following a hurricane in 1998.

CHANGING LANDSCAPES

Any major movement of rock, soil, mud, or snow can change a landscape completely. Forests can be flattened, rivers dammed, fields flooded, and roads and towns destroyed. Even the familiar shape of a mountain can alter when some of its rock suddenly slides away. In 1980, the cone shape of Mount Saint Helens was dramatically altered. An earthquake triggered a landslide that left a vast **crater** at the summit of Mount Saint Helens (see pages 26-27).

LANDSLIDES

WHAT IS A LANDSLIDE?

The term *landslide* covers a wide range of events, from dramatic rockfalls to ground movements that are so gradual they are barely noticeable. The most dangerous landslides occur on steep slopes, for example, on coastal cliffs or in mountain ranges. Even on gentle slopes, such events as torrential rains can set off landslides of slow-moving, loose earth. In the United States, landslides are common in the mountainous regions of the East, for example, the Appalachian Mountains; in the mountainous areas of the West, such as the Rocky Mountains; and in the Pacific coastal ranges, especially in California.

Areas of the United States, particularly mountainous regions, are more vulnerable to landslides than others. Landslide potential is very high in the red areas, high in the yellow, and moderate in the green.

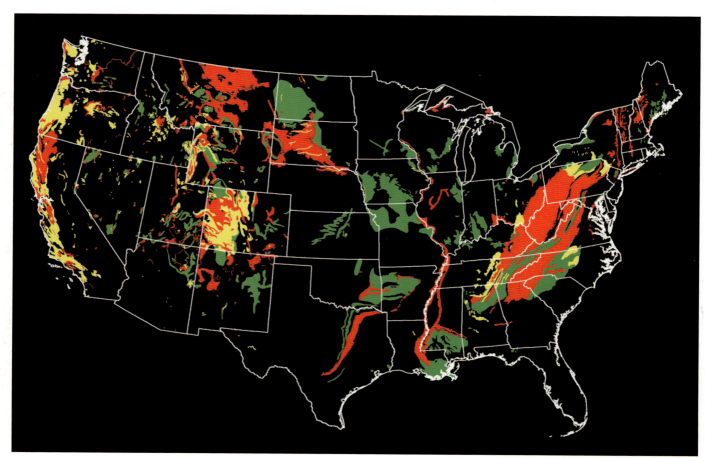

Source: National Atlas and the U.S. Geological Survey.

Fast and slow

The speed of a landslide depends on its make up and inclination (*IHN kluh NAY shuhn*)—that is, the steepness of the angle of the slope down which the landslide is moving. The amount of water, ice, and air contained in the material that makes up a landslide affects how slippery and how heavy it is and whether it moves like a liquid or solid. For example, when water from heavy rain or snowmelt saturates the material of a landslide, filling all the spaces between the particles in the soil, the landslide will tend to move very fast, like a river.

On a steep slope, a landslide can reach speeds of 200 miles (320 kilometers) per hour or more, flattening, smashing, and burying everything in its path. The slowest landslide, on a gentle slope, may move at less than 1 inch (2.5 centimeters) a year. This type of very slow landslide is called **creep** (see pages 8-9).

A giant boulder blocks a road in Topanga Canyon in the Los Angeles metropolitan area. A 25-foot- (7.6-meter-) high boulder came down the mountainside in 2005 as part of a landslide triggered by heavy rains.

TYPES OF LANDSLIDES

The many different types of landslides include falls and **topples,** slides and **slumps,** and **creeps.** They are named according to the type of ground movement and the mix of materials.

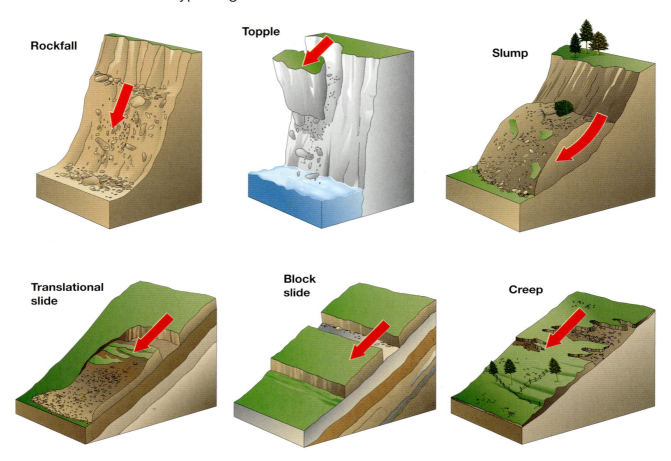

Rockfall

Topple

Slump

Translational slide

Block slide

Creep

The main types of landslides and how they move.

Falls and topples

In places with steep mountains or cliffs, rockfalls present a serious danger. In a rockfall, rocks and boulders plummet down a mountainside at high speeds, often bouncing and crashing off a mountainside or cliff. Rockfalls are a particular risk for climbers and hikers in such places as **canyons.** On May 9, 1999, visitors to Sacred Falls State Park in Hawaii heard a roaring noise that sounded like a freight train. Rock and **debris** fell 500 feet (150 meters) from the top of a canyon wall onto a crowd of people standing at the base of a waterfall in the park. The rockfall killed eight people and injured many others. The park was subsequently closed to visitors.

A topple occurs when material suddenly falls from a vertical rock face, such as a cliff, with forward motion. Like other landslides, topples are potentially dangerous because they are difficult to predict.

Slides and slumps

A slide is a mass of unstable material, such as loose rocks, that moves down over a slope of more stable material, such as a layer of solid rock. There are three kinds of slides, each of which moves in a different way. (1) In a slump, the material moves as one huge block down a concave slope. Because the slope is curved, the material making up the slump tilts backward, instead of falling forward as would occur in a topple. (2) In a translational *(trans LAY shuh nuhl)* slide, the material slides down on a sloping plane. (3) A block slide is made up of just one or a few blocks of material that all slide together. Any of these slides can be fast moving and destructive.

Creeps

A slow, steady landslide of earth or rock is called creep. This type of landslide is usually so slow that the movement is not noticeable to anyone walking on the ground. Over the course of many years, however, walls and trees may begin to lean or crack, because the earth or rock on which they are resting is unstable. Ripples may appear elsewhere on the surface of the ground as the creep continues.

Massive slump-type landslides along the Neelum River in Kashmir were triggered in 2005 by tremors from an earthquake.

LATERAL SPREADS

Some landslides, called **lateral spreads,** occur on land that hardly slopes at all. Unlike other kinds of landslides, lateral spreads can even occur on flat land. A lateral spread happens when rock or earth is resting on a layer of very wet sand or **silt** that suddenly **liquefies** (see pages 12–13). As the lower layer begins to flow like a liquid, the top layer of rock or earth may suddenly break up and begin to spread laterally (sideways) over the surrounding land.

CHINA, 1920

The deadliest landslide event in recorded history took place in 1920, in the provinces of Ningxia and Gansu in north-central China. Landslides, triggered by an **earthquake,** killed around 100,000 people and injured thousands more. The destructive slides affected a massive area of approximately 300 square miles (780 square kilometers). Within this area, 10 cities and many villages were destroyed. The event is remembered locally as the day "when the mountains walked."

Loess landslides

The powerful earthquake that set the landslides in motion occurred along a break in Earth's **crust** known as the Guyuan **fault.** This earthquake had a magnitude of 8.5, and its shaking set off landslides in an area of unstable topsoil called the

The 1920 Guyuan earthquake in China triggered an estimated 600 landslides in four regions of Ningxia and Gansu provinces. The landslides left as many as 100,000 people dead in one of the worst natural disasters in recorded history.

Area covered by the Loess Plateau

Epicenter of 1920 earthquake

Major landslides

Loess (*LOH ihs*) Plateau. **Loess** is a type of **silt** that forms a loose, powdery soil. The earthquake triggered an estimated 600 or more loess landslides. The loess slid down valleys and hillsides at such speeds that few people could escape.

Cave dwellers

Many people who live in this region of China make their homes in caves dug out of cliffs and hills along the river valleys. The earthquake of 1920 sent large sections of cliff falling into the river valleys, destroying hundreds of farms and villages. Many of the cave dwellings collapsed or were blocked by **debris,** trapping people inside.

The loess and debris poured into the rivers, changing their courses. The landslides created at least 40 new lakes in the region as the material dammed the rivers.

MOUNTAINS THAT MOVED

The Gansu catastrophe was described by Upton Close of the International Famine Relief Committee, following a field study in the area: "Mountains that moved in the night; landslides that eddied like waterfalls, crevasses that swallowed houses and camel trains [caravans], and villages that were swept away under a rising sea of loose earth, were a few of the ... occurrences that made the earthquake in Kansu [Gansu] one of the most appalling catastrophes in history. ... In the landslide district ... the loose earth cascaded down the valleys and buried every object in its path."

The landslides associated with the Guyuan earthquake took place in China's Loess Plateau. Loess can easily be carved into cave dwellings and terraced for agriculture but is highly unstable and subject to landslides.

WHAT CAUSES A LANDSLIDE?

Many different events can cause a landslide. The most common triggers are slope **saturation** (SACH *uh* RAY *shuhn*)—for example, from heavy rain or **meltwater—earthquakes,** and **volcanic** activity. Human activity can also be the cause of landslides, if a slope is weakened by excavation or if waste from a mining operation is piled up and creates an unstable slope.

Water

Slopes become saturated with water during heavy rainstorms or when snow and ice melt. When all the spaces between the particles of soil and rock are filled with water, the surplus flows down the slope as **surface runoff.** The weight and movement of the water makes the soil and rock heavy and unstable. The slope fails, and a landslide begins at the point where the force of **gravity** pulling the slope downward becomes greater than the strength of the material making up the slope. Then, the slope itself becomes a **slurry** of water, loose rock, soil, and other **debris** moving downward.

Rescue workers search for bodies after a flash flood caused a fast-flowing slide to sweep through a campsite near San Bernardino, California, on Christmas Day, 2003.

In the United States, **precipitation** is the main trigger for landslides. In December 2003, heavy rains in California led to a deadly landslide. **Wildfires** in previous months burned off the vegetation that had helped to stabilize the slopes in Waterman **Canyon,** near San Bernardino. When heavy rains came on December 25, a sea of mud surged down the slope and claimed the lives of 14 people at a children's camp at the canyon's base. Torrential rainfall was also the cause of devastating landslides in Central America in 1998. When **Hurricane** Mitch hit the region in October of that year, rain fell at a rate of 4 inches (10 centimeters) per hour. Flooding and landslides killed around 10,000 people in the region (see pages 24–29).

Earthquakes

Earthquakes are another major cause of disastrous landslides. During an earthquake, rock moves deep underground, shifting and loosening the rocks and soil above. The jolts and shocks of an earthquake can cause unstable material on a slope to fall as slides, **slumps,** or rockfalls. In Colombia, an earthquake on June 6, 1994, set off landslides in the steep valleys around the Nevado del Huila volcano. The landslides swept downstream and joined in the Paez River valley to form a massive **debris flow** that partially destroyed several towns and killed more than 1,000 people.

If an earthquake strikes in a region where the soil is soft, loose, and wet, solid ground can become unstable though a process called **liquefaction** *(LIHK wuh FAK shuhn).* In such soils, the tiny spaces between individual soil particles are filled with water. As the ground shakes, the movement makes the particles pull apart and float freely, causing the soil to behave like a liquid. On a slope, this liquid material will flow downward, leaving a path of destruction. On flatter land, this process can cause **lateral spreads** (see page 9).

Volcanic activity

Volcanic **eruptions** can produce vast amounts of rock and ash in seconds. This unstable material may slide down a volcano's steep slopes at great speeds. In places where snow lies on a volcano's flanks, the heat from volcanic activity can melt large amounts of snow and ice very quickly. This sudden increase in **meltwater** may increase the speed of the landslide.

CHANGING SLOPES

In many places, slopes that were once covered by trees or other vegetation have been stripped bare. Trees are cut by commercial loggers or to clear space for cropland. Trees and vegetation can also be lost as a result of **wildfires** or **drought.** But once the ground is stripped of its protective cover and left bare, it becomes vulnerable to **erosion,** instability, and in turn, landslide movement.

Illegal logging on mountain slopes in the Philippines has been blamed for several disastrous slides, including floods and landslides that killed around 5,000 people in the city of Ormoc, Leyte Island, in 1999.

WHAT HAPPENS DURING A LANDSLIDE?

The movement of landslides is a complex process, and no two landslides are exactly the same. How the material slides varies, according to the type of material, the slope, and the trigger that sets the slide off. The speed, direction, and makeup of the material often change during a landslide.

People flee a landslide of crashing rocks triggered by heavy rain in the mountains of Kashmir in 2005.

Without warning

Landslides often happen with little or no warning. In mountainous or hilly regions, torrential rain, **earthquake** tremors, or **volcanic** activity may be the only indications that a landslide is about to begin. Both the speed and distance covered by a landslide depend upon the material in the landslide and the **terrain** *(teh RAYN)* over which it is moving. Fast-moving rockfalls may cover a short distance and be finished very quickly. Other landslides may travel many miles, gradually increasing or decreasing in speed according to the steepness of the slope; landslides of this kind may go on for a longer period of time. At Mount Saint Helens (see pages 26–27), the largest landslide in recorded history extended out and around the volcano for 23 square miles (60 square kilometers).

As a landslide moves, it picks up such other materials as trees or **debris** from the buildings it has demolished. The faster the landslide, the more debris it can pick up and carry. Victims of landslides are often found several miles from where they were last seen alive. Some landslides start slowly but then accelerate as the movement of the material gradually causes it to liquefy (see pages 12–13). On Jan. 1, 1997, a landslide in California in the Sierra Nevada mountain range was filmed and its speed measured. The landslide extended for 2.4 miles (4 kilometers). During the first half of its journey, the landslide moved at a walking speed of 2 to 3 miles (3 to 5 kilometers) per hour; during the second half, the landslide speeded up to 12 miles (19 kilometers) per hour. Anyone in the path of the landslide during this second phase would have found escape very difficult.

The end

At some point, when there is no longer a sufficient **gradient** for the material to slide down or when the material has dried out enough to become more stable and less fluid, the landslide stops. A landslide can also come to a halt when it runs up against more rugged land over which it can no longer flow. In some places, the material in a landslide ends up in a river or a lake.

FROM FALL TO FLOW

The characteristics of a landslide can change as it moves. For example, a rockfall may send large blocks of rock crashing down a mountainside. As the rocks bounce and fall, they break into increasingly smaller fragments. If these fragments mix with water, they may quickly form a fast-flowing **debris flow** (see page 24). With the addition of smaller particles from lower slopes, such a flow may pick up speed and travel distances of 60 miles (100 kilometers) or more from the original fall.

The foot of a landslide in Berner Oberland, Switzerland, spills out into a river. Debris from landslides can pollute river water and endanger wildlife.

LANDSLIDE EFFECTS

Landslides have an impact on **ecosystems,** especially on the wildlife that lives within, as well as on human settlements. Like other disasters, landslides can cause loss of life, injury, and homelessness. Emergency aid, clean up, and rebuilding all have long-term financial consequences too.

Changing landscapes

Large landslides can drastically change a landscape. Massive mountainsides can disappear in seconds, and whole villages can be swallowed up. Some landslides are big enough to create lakes by damming rivers. In 1983, heavy autumn rains, followed by heavy winter snows and a rapid spring melt, led to a landslide that dammed the Spanish Fork River, near the town of Thistle, Utah. The water held back by the "dam" formed a lake that inundated the town. Within 24 hours, the town's buildings had disappeared beneath the floodwaters. Luckily, the town's inhabitants had been evacuated. The amount of damage caused by this landslide was estimated at more than $600 million.

A vast landslide (the dirty white in an aerial view) dams a river near Thistle, Utah, in 1993, creating a lake that rose to cover the town within 24 hours.

Collapsing buildings

The tons of rocks and other material contained in a landslide can easily crush buildings. When material begins to slide, any buildings that sit on top of it soon collapse as their foundations give way. If a landslide reaches a place where there are many buildings, such as a town or a city, large numbers of people are likely to be killed, injured, or made homeless. The 1994 landslide in the Paez River valley in Colombia (see pages 12–13) left thousands without shelter.

Disruption of services

The movement of rock and earth can destroy pipes and wires, cutting people off from clean water, electric power, and telecommunications. Landslides can block roads and make it difficult for emergency services to reach trapped or injured people.

In June 2001, torrential rain caused a series of disastrous landslides in Ecuador. One of these slides, near the town of Papallacta, destroyed a 200-foot (60-meter) section of Ecuador's main oil pipeline and caused large amounts of crude oil to be spilled.

Cleanup

Oil spills, such as the one in Ecuador, can have a devastating impact on the local environment, polluting rivers and the surrounding land and killing wildlife. Leakage from broken oil or **sewage** pipes can also lead to the pollution of clean water supplies. Once all efforts have been made to rescue landslide victims, the next job is to provide fresh water and shelter. Afterward, the long-term work of cleaning up and rebuilding begins.

Rescue workers carry a survivor from the 2001 slide near Papallacta, Ecuador, to safety and medical aid.

OUTBREAKS OF DISEASE

Landslides can sometimes result in the outbreak of disease. In 1994, landslides caused by the Northridge earthquake in California released large amounts of dust into the atmosphere. These dust clouds gathered over the Simi Valley, near Los Angeles. The dust contained fungal spores (single cells, produced by **fungi** [FUHN jy], that are able to reproduce). These spores, when breathed in, caused a disease found in the western United States, called valley fever, or coccidioidomycosis (kok SIHD ee OY doh my KOH sihs). After the landslides, an epidemic of this fungal infection occurred, with more than 150 people diagnosed with valley fever. Three people died in this outbreak.

ALASKA, 1964

On March 27, 1964, a huge **earthquake** in Prince William Sound, Alaska, set off numerous landslides, mainly caused by **liquefaction** of the soil. The earthquake had a magnitude of 9.2, and its **epicenter** was about 75 miles (120 kilometers) off the Alaskan coast. Many towns and cities were damaged by the earthquake and ensuing landslides, including Anchorage, Seward, Valdez, and Whittier.

Hillside homes lie in ruins on Turnagain Heights in Anchorage, Alaska, after the 1964 earthquake triggered devastating landslides. At least 75 houses in the neighborhood were destroyed.

Sliding buildings

The 1964 earthquake was the biggest ever recorded in North America, and the ground in the Anchorage area was reported to have moved for between three and four minutes. The violent shaking damaged or destroyed many buildings and also triggered landslides as a result of the liquefaction of an underground layer of clay. In Anchorage, the landslides caused enormous damage. The Government Hill Grade School, for example, sat directly above a huge area of landslide and was almost completely destroyed.

The most destructive landslide in Anchorage, in Turnagain Heights, devastated an area of 130 acres (52 hectares). Seventy-five houses and many other buildings tilted or collapsed when a steep bank fell in a **slump**-type landslide. Service pipes, power lines, and telephone wires were also damaged.

Landslides at the towns of Seward and Valdez shifted their waterfronts

into the sea, forever changing the shape of the coast. In Seward, the landslide left train tracks running directly into the Gulf of Alaska.

Tsunami

The movement of the seafloor during an earthquake can push the surrounding seawater into a series of large, powerful waves, called **tsunamis.** During the 1964 earthquake, **uplift** and landslides on the ocean floor caused five tsunami waves to strike the Alaskan coast within minutes. The waves, measuring up to 220 feet (67 meters), damaged towns all along the Gulf of Alaska.

EYEWITNESS: TSUNAMI

Peggy and Roxy Coons, keepers of the Battery Point Lighthouse in Crescent City, Alaska, described a tsunami wave caused by the 1964 underwater landslides: "The water withdrew as if someone had pulled a plug. ... Then the mammoth wall of water came barreling towards us. It was a terrifying mass. ... It struck with such force and speed that we felt like we were being carried along the ocean."

Source: *The Raging Sea* by Dennis Powers.

Sections of the Alaskan railroad disappear into the sea at Seward, where the 1964 earthquake triggered a landslide that shifted the waterfront out into the Gulf of Alaska.

AVOIDING LANDSLIDES

The best way to avoid landslide fatalities and building damage is to prevent people from building on sites with a high risk of landslides. **Geologists** research and advise on which areas are vulnerable to landslides. Local authorities often post warning signs to make passersby aware of the dangers in these unstable areas. Government authorities also try to enforce regulations restricting the building of homes and other structures on dangerous sites. This can be difficult, however, as land on coasts and in mountains is often desirable for development.

A sign near the Karakoram Highway in Pakistan warns travelers to beware of landslides.

Stabilizing slopes

There are various methods of stabilizing slopes to reduce the risk of landslides. In places where large amounts of water run down a slope, channels and pipes can be put in place to carry the water away safely. Placing a waterproof covering onto the loose material covering the slope can also help to protect it. Such coverings prevent a slope from becoming **saturated** and unstable. In places where loose material builds up at the top of a slope, moving the material to the bottom can increase stability and make a landslide less likely. If such measures as these are not taken, disaster can follow. In October 1966, heavy rain saturated tons of coal mine waste, called slag, that had been tipped on a mountainside above the village of Aberfan, in the United Kingdom. The slag had been piled up over a period of 50 years. Rainwater mixed with the slag to turn it into a heavy black sludge that suddenly slid down the mountainside. The landslide buried 20 houses and a school in Aberfan. Most of the 144 people killed were children.

Residents of the Welsh mining village of Aberfan search for victims buried by the catastrophic landslide of 1966.

Evacuation and preparation

In places where landslides are likely, scientists can watch for potential triggers, such as heavy rain, so that people nearby can be evacuated before a disaster occurs. Although evacuation does not save buildings, it does save lives. On Sept. 24, 2003, landslides fell from a Himalayan mountainside onto the township of Uttarkashi, in northern India. However, the area had already been evacuated, following warnings from geologists who had detected some movement on the slopes after heavy rains.

LANDSLIDE PRECAUTIONS

The United States Geological Survey (USGS) recommends the following precautions:

- Stay awake during an intense storm in an area with a likelihood of landslide. Many landslide fatalities occur when people are sleeping. Listen to the radio or television for warnings during periods of intense rainfall.

- During a time of heavy rainfall, consider leaving an area susceptible to landslides if it is safe to do so. Getting out of the path of a landslide is your best protection. If you decide to remain at home, move to a second story, when possible.

- Listen for sounds, such as cracking trees or shifting boulders, which might indicate moving **debris.**

- If you are near flowing water, be alert for any sudden increase or decrease in water flow or a change from clear to muddy water. These can be signs of landslide activity upstream, so be prepared to move quickly if you see these things.

- Be cautious when traveling in a car during intense storms. Do not cross flooded streams. Watch the road for collapsed pavement, mud, fallen rocks, and other indications of a possible landslide.

SUBMARINE LANDSLIDES

The ocean floor is not flat. It has deep valleys and high mountains, just like those that we see on land. The ocean floor around the continents has areas of higher ground, called **continental shelves.**

Earthquakes and **volcanic** activity can trigger underwater landslides, just as they can on land. Landslides can occur where the continental shelves slope down to the deeper ocean floor. These landslides, and those on other underwater slopes, are called submarine landslides.

Scientists traveling in a submersible examine and map the ocean floor. Submersibles allow scientists to observe phenomena at great depths.

Landslide maps

Scientists use **sonar** instruments and **submersibles,** as well as images taken by **satellites,** to map the ocean floor. Information from these sources has revealed massive piles of **sediment** on the ocean floor—evidence of submarine landslides that have occurred in the past. For example, in 2000, scientists studying the coastline off Santa Barbara in California found evidence of many submarine landslides in the area. One pile of sediment was 9 miles (14.5 kilometers) long and 6.5 miles (10.5 kilometers) wide. From samples taken from the ocean floor, scientists are now trying to work out exactly when and how these huge landslides occurred.

Scientists believe that some underwater landslides are far bigger than any landslide recorded on land. For example, near Ruatoria, off the coast of North Island, New Zealand, there is evidence that a vast area of ocean floor, measuring roughly 25 by 50 miles (40 by 80 kilometers), fell down an underwater **scarp** 2 miles (3 kilometers) high. The massive rockslide traveled 30 miles (50 kilometers), spreading out across the flatter ocean floor below. Among the fallen rocks and mud were individual rocks the size of mountains.

An artist's impression of a huge underwater landslide.

SLIDES AND WAVES

Some evidence indicates that submarine landslides can trigger a **tsunami.** In July 1998, a tsunami occurred in the Pacific Ocean along the north-central coast of New Guinea. The tsunami had waves between 33 and 50 feet (10 and 15 meters) high, which roared onto the coast and killed around 2,500 people. The back surge from the tsunami was so powerful that bodies were found washed up on beaches in Indonesia, some 25 miles (40 kilometers) away. Initially, scientists sent to survey the site were surprised by the intensity of the tsunami, which followed a fairly moderate earthquake with a magnitude of 7.1. They eventually theorized that it was not an earthquake that caused the tsunami, but an underwater landslide triggered by the earthquake.

MUDSLIDES

WHAT IS A MUDSLIDE?

The word *mudslide* does not have a precise scientific meaning. While the term is often used to describe a rapidly moving landslide of very wet material, in different regions the word *mudslide* can be used to describe different types of events.

Flowing mudslides

In some areas, such as Nepal and the Philippines, mudslide describes a flowing, liquid "river" of water and mud. Mudslides such as these can be so powerful that large objects—boulders, houses, trees, and people—may be picked up and carried in the flow for miles. Heavy rains are nearly always the cause of these fast-moving rivers of mud. As the sand, **silt,** or clay on a slope becomes **saturated** with rainwater, the slope becomes unstable and begins to move downward, starting a mudslide. Mudslides are a seasonal problem in some regions because they tend to occur during the heavy rains brought by **monsoons.** Torrential rain during **hurricanes** and other storms and **meltwater** in mountainous regions can also trigger mudslides. Finally, flooding can be a trigger for mudslides. If a lake or river on high ground overflows its banks, the floodwater will stream downhill. As the floodwater mixes with sand and silt along its way, it can soon become a raging river of mud.

A worker repairs broken electric power lines in the Philippines after heavy rains in 2006 triggered a mudslide that flowed from Kan-Abag Mountain (in the background) to bury the farming village of Guinsaugon. More than half of the village's 1,800 residents were killed.

Volcanic mudslides

The term *mudslide* is also used to describe a **lahar,** a liquid mudflow made up of water, **volcanic** ash, and **debris.** Lahars are often associated with volcanic **eruptions.** The heat of an eruption can melt snow and ice on a volcano, and this meltwater mixes with volcanic ash and debris to form a flowing river of mud. Heavy rains can also create a lahar. In 1998, unusually heavy rains triggered a lahar on Casita Volcano in Nicaragua, which killed 2,500 people as it streamed down the mountainside.

A house is partially buried and surrounded by debris from a lahar that flowed from Mount Unzen in Japan after the volcano erupted in 1991.

Rain-induced landslides

Finally, *mudslide* can sometimes be used to describe an event that should properly be termed a landslide, such as when a large section of cliff or hill gives way and comes crashing down a slope. These events often happen in coastal California. Heavy rains are nearly always the trigger for these occurrences.

Little time to escape

Flowing mudslides and lahars move very rapidly on steeper slopes, so there is often little time to escape from them. Typical speeds are between 10 and 35 miles (16 and 56 kilometers) per hour, but mudslide speeds of more than 100 miles (160 kilometers) per hour have been recorded. Survivors of the devastating mudslides in Central America during Hurricane Mitch in 1998 reported experiencing torrential rain followed by a noise that sounded like "a fleet of helicopters ... Within minutes an avalanche of mud, tree trunks, and rocks wiped out everything."

MUDSLIDES IN THE PHILIPPINES

Heavy rain and tropical storms frequently trigger devastating floods and mudslides in the Philippines. In 1991, around 6,000 people were killed on the southern island of Leyte in mudslides and floods following a tropical storm. On Feb. 17, 2006, disaster struck again when 10 days of heavy rain set off several mudslides. The mountain village of Guinsaugon was completely destroyed, with nearly all of its buildings buried in mud. The estimated death toll was well over 1,000 people. Another tropical storm hit the Philippines in December 2006. The heavy rains triggered mudslides southeast of Manila, claiming at least 400 victims.

DEVASTATING VOLCANIC MUDSLIDES

Volcanic mudslides can cover large areas. If a fast-moving mudslide hits a settlement, the loss of life can be devastating. That was the case in 1985, when the town of Armero in Colombia disappeared beneath a river of mud. Mudslides can also devastate the land. The biggest landslide in recorded history, at Mount Saint Helens in 1980, caused a mudslide that ravaged a large area. Fortunately, that area had already been evacuated, so the loss of life was low.

The destruction of Armero

Nevado del Ruiz is a snow-covered **volcano** that rises 17,717 feet (5,400 meters) in the Andes Mountains in Colombia. Although only around 300 miles (480 kilometers) from the equator, the volcano's height causes its summit to be covered in snow and ice year-round. On Nov. 13, 1985, Nevado del Ruiz erupted, sending hot rock, **pumice,** and ash high into the atmosphere and rapidly melting large amounts of snow and ice on the volcano's summit. Mixtures of **meltwater,** ice, and rock began to pour off the summit, in some places carving channels that were from 6 to 12 feet (2 to 4 meters) deep. After descending several thousand feet, the fast-moving **lahars** were channeled into river valleys leading from the volcano.

The lahars raced along the valleys of the Chinchiná, Gualí, and Lagunillas rivers at average speeds of 37 miles (60 kilometers) per hour. As they poured downhill, they killed more than 20,000 people. Worst hit was the town of Armero, on the Lagunillas River. Two hours after the **eruption,** a lahar almost completely destroyed the town.

The remains of Armero, Colombia, lie buried by a lahar two hours after the eruption of Nevado del Ruiz on Nov. 13, 1985.

Mount Saint Helens

On May 18, 1980, a small **earthquake** on Mount Saint Helens sent about 3.7 billion cubic yards (2.8 billion cubic meters) of rock slipping down the north side of the volcano—enough material to cover a large city. When this side of the mountain gave way, it triggered a lateral (sideways) eruption. Hot gases, ash, rocks, and steam spewed out of the volcano, melting the snow and ice on the slopes. The meltwater and loose **debris** quickly formed lahars.

One of the lahars flowed into the North Fork Toutle River and grew rapidly as it was fed by water and **sediment** from the river. As the lahar traveled along the river's course, it destroyed bridges, roads, railways, and buildings. After the lahars had stopped, settled, and dried, they left deposits that covered a huge area of many square miles.

Despite the widespread destruction, only 57 people died as a result of the eruption of Mount Saint Helens, most of them from inhaling hot volcanic ash. The area surrounding Mount Saint Helens had been evacuated in anticipation of the eruption.

A lahar, clearly visible as a dark deposit on the white snow, flows down Mount Saint Helens in 1980. Part of the lahar continued into Spirit Lake (bottom left) while the rest flowed west to the North Fork Toutle River.

MOUNT SAINT HELENS STATISTICS

- Only 14 of 32 known **species** of small mammals in the area survived the eruption and its aftermath.

- The lateral blast triggered by the landslide completely blew down, scorched, and killed all trees within a 15-mile (24-kilometer) radius north of the volcano.

- The speed at which gases and rock were released in the eruption blast ranged from 220 to 670 miles (350 to 1080 kilometers) per hour.

- The overall cost of the damage and destruction caused by the 1980 eruption was originally estimated to be from $2 billion to $3 billion. It was later revised to $1.1 billion, still a huge amount of damage.

RAIN-INDUCED LANDSLIDES

One type of slide, though called a mudslide, is actually a landslide. Common in coastal areas of California, they generally consist of large sides of a hill or cliff that **slump** off and rush down a slope. They are caused by heavy rains.

La Conchita

The town of La Conchita on the coast of southern California has a history of landslides dating back as far as 1865. The town lies on a narrow coastal strip with the sea on one side and a hillside rising about 600 feet (180 meters) on the other. The hillside is made up of such loose materials as siltstone, sandstone, and mudstone.

A cliff-top view of the path of the mudslide that destroyed or buried 24 buildings, killing at least 10 people, in La Conchita, California, in January 2005.

In January 1995, over 18 inches (46 centimeters) of rain—six times the normal amount for that period—fell in the La Conchita area. A storm on March 2–3 dropped more heavy rains on the already wet soils, making the slopes of the hillside behind La Conchita unstable. On March 4, around 1.7 million cubic yards (1.3 million cubic meters) of earth slid down, and within a few minutes the slide had destroyed and damaged several houses. The mudslide covered 10 acres (4 hectares). On March 10, a **debris flow** from a nearby **canyon** reached La Conchita too, damaging five more houses. Fortunately no one was killed.

Another serious mudslide occurred at La Conchita on Jan. 10, 2005. The mudslide came after 15 days of heavy rainfall. Residents had already been evacuated, though 18 people refused to leave their homes. The mudslide reached farther into the town than the 1995 mudslide, destroying 13 houses and damaging 23 more. Rescuers spent two days looking for missing persons, and 10 people are known to have died. Following this event, local Sheriff Bob Brooks noted: "The La Conchita community is a geologically hazardous area. ... We do not recommend that people return to this area." Nevertheless, many of the residents of La Conchita did return.

Rescue workers search through the wreckage of one of the La Conchita houses destroyed by a mudslide on Jan. 10, 2005.

South Canyon wildfire

Wildfires are fairly common in California. A single fire can spread quickly and destroy vegetation across large areas of countryside. Plants help to hold soils in place, so the removal of vegetation makes exposed soils more vulnerable to slippage. In 1994, a wildfire in South Canyon, California, was followed by heavy rain. The rain quickly turned the exposed clay soil on the burned slopes into a **debris flow.** The fast-moving flow swept across Interstate 70 and into the Colorado River, nearly damming the river.

LANDSLIDE INFORMATION

In such areas as California, where the risk of landslides and mudslides is high, maps are available from the United States Geological Survey (USGS) to help people make decisions about where to build safely and where to take precautions:

- Susceptibility maps show where landslides may occur by ranking slope stability in a given area.

- Hazard maps indicate the possibility of landslides occurring. They may show the locations of past landslides as well as more complex information about the makeup of the earth and rock, slope angles, and **earthquake** tremor measurements.

- Risk maps show the cost of landslide damage in a given area.

AVALANCHES

WHAT IS AN AVALANCHE?

An avalanche, or snow slide, is a mass of snow, or sometimes snow and ice, that slides down a mountain slope. (The word avalanche is also sometimes imprecisely used to describe a rockfall.) Avalanches are a major danger in mountainous regions. Snow avalanches can be large enough to bury a village and can have enough force to pull boulders and other **debris** along with them.

A slope with a slant of less than 25 degrees is so flat that large amounts of snow are unlikely to slide down it. Snow slides so quickly down a slope steeper than around 50 degrees that it cannot usually accumulate to a great enough depth for an avalanche to occur. That is why avalanches usually occur on slopes with a slant of from 25 to 50 degrees—the kind of slopes that are, in fact, good for skiing. No matter how steep or flat the slope, however, snow will not suddenly form into an avalanche unless the snow has become unstable. As with a landslide or mudslide, **gravity** is the force that pulls unstable snow downward.

Behind a powdery cloud, tons of snow rumble down a steep Himalayan mountain slope in an avalanche near Mount Everest, in Nepal.

Sluffs and slabs

There are two main types of avalanche: a sluff avalanche and a slab avalanche. A sluff avalanche may be made up of wet or dry snow. A sluff avalanche of dry snow is made up of powdery snow and air. This type of avalanche can move faster than 100 miles (160 kilometers) per hour. Powdery snow occurs in very low temperatures because in very cold weather, snow crystals are less likely to stick together. A sluff avalanche of wet snow consists of wet, dense snow and usually moves more slowly. Snow becomes wet and **compact** when higher temperatures make the ice crystals melt and stick together.

In a slab avalanche, a solid section of snow breaks loose from the surrounding snow. The sides of the slab then split into pieces. Solid slabs of snow often form when a sunny, warm day is followed by a cold night. The snow melts a little during the day; at night the **meltwater** freezes and sticks the remaining snow back together in a solid slab. Most avalanche fatalities occur during a slab avalanche.

WHO IS IN DANGER?

The countries with the highest risk of avalanche are Austria, France, Italy, Switzerland, and the United States (especially the states of Alaska, Colorado, and Utah). Millions of tourists visit such mountainous areas as the Alps and the Rockies every year, increasing the number of people in danger when avalanches do occur. Further, the practice of skiing, snowboarding, and snowmobiling in backcountry areas—areas away from monitored and maintained trails—has increased the number of avalanche fatalities.

A skier explores a rugged landscape of snow blocks after a 2002 slab avalanche in Sweden.

Signs warn drivers that a mountain road in Washington state is closed because of avalanches. As winter snow melts, avalanches become more frequent, and road closures are often necessary.

WHY DO AVALANCHES HAPPEN?

Most avalanches are caused by such changes in the weather as a sudden, heavy snowfall or a rise in temperature that makes the snow begin to melt. Such weather changes make snow on a mountain unstable and more liable to move. Disturbances that can set the snow moving include heavy winds, tremors from **earthquakes,** and explosions. People moving over an unstable area, such as skiers or snowmobilers, can also set off an avalanche.

Snow layers

Snow layers are an important factor in avalanches. Snow builds up in layers if it does not melt between snowfalls. This layering can lead to avalanches in several different ways. Sometimes, the weight of new snow makes lower layers become **compact,** like ice. Eventually, these compacted layers may become so heavy that the force of **gravity** pulls them down in an avalanche. In other instances, new powdery snow that falls onto a smooth layer of compacted snow is likely to be unstable, and this unstable snow can suddenly slip off the lower layers, forming an avalanche. The risk of avalanche also increases if the new snow is above the level of vegetation or rocks that can help to hold it in place. Further, sometimes when snow starts to melt, the lower snow layers become less "sticky," as the **meltwater** runs down, dissolving the bonds between the snow grains. In this way, the lower snow layers decrease in strength and stability and may begin to move as an avalanche. Rushing meltwater may increase this effect, pushing remaining snow along with it as it pours down slopes.

More snow

The risk of avalanche increases when there is high snowfall. In the winter of 2007, three snowstorms in three weeks piled up to 4 feet (1.2 meters) of snow in some parts of Colorado. The unusually heavy snowfall led to a large avalanche on January 6, which swept onto a highway, knocking several cars off the road and burying them near the ski resort of Winter Park. Fortunately, people trapped in the cars were quickly rescued.

Winds

The force of high winds can set off an avalanche. Winds can also push snow into huge drifts on slopes. These thicker snow layers may become so heavy they fall as an avalanche. Winds bring changes of temperature too. In the Alps, warm, violent winds called foehns *(faynz)* blow downward along mountain slopes. The winds melt snow and ice on the mountainsides, setting off avalanches.

Vibrations

An earthquake can suddenly loosen snow and cause catastrophic avalanches. Explosions can also set off an avalanche. Often, in fact, experts deliberately set off a small avalanche using explosives (see pages 42-43). These smaller, controlled avalanches prevent larger avalanches from occurring.

WINTER SPORTS

Most of the people who die in an avalanche while pursuing such winter sports as skiing or climbing have triggered the avalanche themselves. In places where the snow is unstable, a person's body weight can set an avalanche in motion. Surprisingly, although skiers can trigger an avalanche, only a very small number of avalanche deaths occur at ski resorts. The reason is that the slopes at a resort are highly maintained and made safer. When too much snow builds up, explosives are used to cause a mini-avalanche. Such measures keep ski resorts relatively safe from major avalanches. It is the areas away from the ski resorts that have proved deadly in recent times. Around 75 percent of avalanche victims today are backcountry recreationalists using more extreme wilderness areas in which to hike and ski.

A skier attempts to outrun an avalanche in Utah.

THE ALPS, 1999

The year 1999 was a very bad year for avalanches in the Alps—the worst in around 50 years. The amount of snow that fell in the region late in 1998 and early in 1999 was unusually heavy. At one point, the snowfall was so intense that food and supplies had to be airlifted to thousands of tourists who were trapped for days in various **Alpine** resorts. The heavy accumulation of snow, coupled with conditions that allowed the snow to repeatedly thaw and refreeze, created a season of many avalanches. Around 1,000 avalanches killed 86 people in the Alps in early 1999. In February 1999, two avalanches killed 38 people in the Austrian Alps in Galtür and the nearby village of Valzur (see page 36); an avalanche killed 12 people in the French Alps, near Chamonix; and yet another avalanche killed 12 people in the Swiss Alps, near the village of Evolene.

The danger from avalanches was possibly compounded by the increase in building that had occurred over the past decades in mountainous areas of the Alps. Small chalets dotted the landscape. As the pressure increased to develop more land, more structures were located in the natural paths that avalanches followed in the past and were likely to follow in the future.

A crushed car reveals the power of a Feb. 9, 1999, avalanche that killed 12 people in their homes in the French Alpine village of Montroc, near Chamonix.

Powder cloud

The wall of snow that travels down a slope during an avalanche is frightening. But the powder cloud of an avalanche is also a major force. In a large avalanche, a cloud of swirling, fast-moving snow crystals can travel above the moving snow. This cloud usually travels faster than the snow slide part of the avalanche. The force of this cloud is such that it can snap trees and move cars. When the snow portion of an avalanche hits an uphill slope, it stops. But the powder cloud can travel for miles, up and down slopes. Just before the avalanche hit at Galtür, Austria, survivors reported that the sky went dark. The large powder cloud of this avalanche had obscured the sun.

SWISS EXPERTISE

Switzerland has one of the most advanced warning and prevention systems for avalanches of any nation in the world. The Swiss Federal Institute for Snow and Avalanche Research at Davos is responsible for assessing the risk of avalanche and for warning the population about possible avalanches. During the peak of avalanche season, the group produces the Avalanche Bulletin, which is meant to help local forecasters. About 100 bulletins are published over the course of one winter.

Other measures taken to combat avalanches in Switzerland include:

- Avalanche barriers—steel posts, fences, and even dams, which are meant to prevent large amounts of snow from moving over layers of previously fallen snow
- Avalanche hazard maps—using historical avalanche routes to try to predict the routes a new avalanche might take
- Avalanche building—houses and buildings reinforced to withstand avalanche snows
- Avalanche observation—a national avalanche information system using both human observers and remote weather stations

During the summer months it is easy to see the avalanche barriers on a steep mountain slope in Switzerland. In the winter, these barriers help to prevent snow from sliding down into the valley below.

WHAT HAPPENS DURING AN AVALANCHE?

An avalanche can appear as a cloud of airborne crystals, as massive sliding chunks of snow, or as a huge slab of snow. Once an avalanche has started, it moves so fast that it is usually impossible for a person to outrun or outski it. Some avalanches reach speeds of 200 miles (320 kilometers) per hour. The speed of an avalanche can vary as it falls, and it may increase or decrease in size as well.

Airborne and booming

In an airborne, or wind, avalanche, dry, powdery snow is blown along in the air, creating a cloud of snow. As the cloud goes down a slope, the air pressure builds up in front of it. Avalanche survivors often experience this booming blast of air before the snow hits them. In 1951, a man in Andermatt, Switzerland, was blown 100 feet (30 meters) in less than a minute, as was the roof of his house. Next, he

Rescue workers and volunteers search for survivors after the avalanche that partially buried the village of Galtür, Austria, in 1999.

found himself buried by the avalanche that followed. He survived to find his house buried under 30 feet (9 meters) of snow.

Smaller or bigger

Avalanches often change in size as they progress. A slab avalanche may start as a huge block, or series of blocks, but these often break up as they move to form smaller avalanches. However, this was not the case with the avalanche that hit Galtür, Austria, in 1999. At the starting zone of the avalanche, an area 1,600 feet (490 meters) wide and 10 feet (3 meters) deep began to slide. In the 50 seconds the avalanche took to come down its track (avalanche route), it doubled in size by picking up more snow along the way. By the time it hit the village of Galtür, the avalanche appeared as a wall of snow and ice 300 feet (90 meters) high.

Residents dug 20 survivors out of the snow, but by nightfall 31 people were still missing. Helicopters and rescue teams had not reached the town. The next afternoon, a second avalanche hit the nearby village of Valzur. The following day, 65 helicopters were used to evacuate nearly 20,000 people in the area. In all, 38 people died.

SNOWMOBILE RISKS

In the United States, most people killed in avalanches are on snowmobiles. The majority of snowmobile avalanche deaths occur while riders are taking part in a sport known as high marking—trying to reach the highest spot on a slope the snowmobile is capable of, then heading quickly down the hill once that point is reached. Often, the weight of the snowmobile and rider on an unstable slope can lead to disaster.

Winds also build up snow into unstable crests that daredevil riders attempt to jump off. However, such crests often give way, and riders can find themselves falling in an avalanche down the slope.

The weight of a snowmobile and rider can cause unstable snow to collapse, triggering an avalanche.

A massive avalanche crashes through an Alpine forest in Switzerland, ripping up trees and burying much of the animal and plant life.

THE EFFECTS OF AVALANCHES

Avalanches have many environmental effects. The force of the snow itself can cause enormous damage to vegetation and animals. Rocks and other **debris** moved by the power of the snow are displaced, further affecting landscapes. However, environments do slowly recover, and some can actually benefit from the **meltwater.**

Mountain forests

A single large avalanche can rip up trees in mountain forests, bury animals, and carry boulders down a slope, causing widespread habitat destruction. Smaller seasonal avalanches often follow the same route. In mountain forests, they can leave a treeless path that becomes wider with each avalanche.

New growth and wildlife

Patches or pathways cleared by avalanches provide open habitats among the closed mountain forests, helping to increase the diversity of wildlife in the area. In parts of the Rocky Mountains, for example, scientists have observed a rich mix of forest and meadow **species** at boundaries between mountain forests and areas cleared by avalanches. Flowering and fruiting shrubs thrive, providing food for birds and such animals as grizzly bears. Paths cleared by avalanches also attract birds of prey. In the Rockies, golden eagles hunt for marmots, ground squirrels, and other small mammals over such areas. As winter snow melts, golden eagles and grizzly bears search out the frozen remains of dead animals, such as goats or elk, that had been killed in avalanches.

HUMAN FATALITIES

Because some snow is light and powdery, it would appear that it would be fairly easy for people caught in an avalanche to dig themselves out once the snow has stopped moving. But that is not so. People are often knocked unconscious by being tumbled in the mass of snow or by striking objects on the way down the slope. And even if they remain conscious, once the snow has stopped sliding, they may be unable to tell which way is up. Finally, once an avalanche has stopped, after three to four seconds, the snow hardens to something like cement. People buried in an avalanche of snow only up to their knees have sometimes needed to have rescuers dig them out. The snow was packed so hard they were unable to free themselves.

Lower slopes

At the run-out zone (where the avalanche comes to a halt), anything that is being carried in the avalanche, such as boulders, rocks, and other debris, is then dropped. These deposits left behind after the snow has melted can affect river routes and drainage in the area. The snow may also carry **nutrients** scoured from the slope down to lower regions where plant life will benefit.

When snow at the bottom of an avalanche pathway melts, it provides an additional source of water that increases the range of plant growth. However, very thick deposits of avalanche snow may take most of the summer to melt, suppressing plant growth until late summer.

A trail of snow and ice left by an avalanche slowly melts, providing extra water for plant growth.

STEVENS PASS, 1910

In the United States, an average of 20 to 30 people are killed in avalanches every year, and avalanche-related deaths occur in every month of the year. While most avalanches take place between December and April, mountaineers are sometimes the victims of avalanches at other times, because it remains cold year-round at high **altitudes.** The worst snow avalanche disaster in the United States, which occurred during the winter of 1910, killed 96 people.

Trapped trains

In February 1910, two westbound trains on the Great Northern line were delayed in central Washington state for several days while snow was cleared from the tracks. They were heading into the Cascade Range of mountains. When the trains finally set off, they were stopped by avalanches 60 miles (97 kilometers) northeast of Seattle, on the railroad tracks at Stevens Pass (altitude 4,016 feet [1,224 meters]). The trains and passengers were stuck on the pass for six more days.

The weather worsened, with blizzards that increased the likelihood of additional avalanches. On March 1, shortly after midnight, a thunderstorm triggered an avalanche on nearby Windy Mountain. Sweeping down the mountainside, the avalanche pushed the locomotives and railroad cars off the track and over the mountain ledge.

The wreckage of one of the two locomotives that were swept off the railroad track and down 1,000 feet (300 meters) into a creek by the Stevens Pass avalanche in 1910.

The trains rolled 1,000 feet (300 meters) down into Tye Creek, where they were buried by 40 feet (12 meters) of snow, rock, and broken trees.

Rescue

Residents of nearby Wellington were the first on the scene. It took six hours to find the wrecked trains. Some of the lucky survivors had been thrown from the trains. Others had to be dug from the snow. In all, only 23 people were rescued. Bodies were still being discovered three months later. Of the 96 dead, 35 were passengers and 61 were railroad workers.

EYEWITNESS

One of the survivors recalled the disaster: "Lightning flashes were vivid and a tearing wind was howling down the canyon. Suddenly there was a dull roar, and the sleeping men and women felt the passenger coaches lifted and borne along."

Local residents search for survivors in the tangled wreckage of the railroad cars after the Stevens Pass avalanche of 1910.

AVALANCHE PREVENTION AND RESCUE

The best way to reduce fatalities and damage from avalanches is to try to prevent them from happening. The first priority is to identify potential avalanche sites. Then various methods can be employed to try to prevent avalanches or to limit their damage.

Anticipation

Experts base predictions of the likelihood of avalanches in an area by studying its **terrain,** snow, and weather. Such organizations as the American Avalanche Association work to study and forecast avalanches. Avalanche centers in various countries and states (including Colorado, Idaho, and New Hampshire) provide forecasts so that skiers, mountaineers, and residents can avoid avalanche sites. Forecasts may also lead to the evacuation of an area.

Avalanche defenses

Explosive charges may be placed on slopes where large amounts of snow frequently accumulate. In winter, the explosives are set off to shift the snow in small, safe avalanches before the snow builds to a dangerous level.

To protect buildings and villages, trees may be planted to slow down any avalanches on nearby slopes. Fences and walls may also be built as snow barriers. Curved walls or earth mounds are sometimes constructed to divert potential avalanches away from groups of buildings.

During heavy snow in 2002, rangers at the Alta Ski Resort in Utah prepare to fire a rifle to stop the unstable layering of snow.

Rescue teams and survival

Many ski resorts have rescue teams in case of avalanches. Specially trained dogs are used to pinpoint the location of avalanche victims. Skiers are also advised to wear special transmitters that, if buried in an avalanche, will send out signals that can be picked up by rescuers. Rescue teams use long, thin probes to locate the exact position of people buried in the snow and strong shovels to dig them out. Most avalanche victims have a good chance of survival if they are rescued within 15 minutes of being buried, but their chances decline rapidly beyond that length of time, so fast response and rescue is vital.

A person who finds him- or herself in an avalanche is advised to use swimming and rolling motions to try to stay as near the top of the snow as possible. Once the snow stops moving, it sets hard within seconds. Chances of survival are improved if the victim manages to create an air space in front of the mouth and nose and to expand the chest in order to have a small amount of breathing space as the snow solidifies. Most avalanche deaths occur from suffocation (lack of oxygen for breathing) or from poisoning by carbon dioxide—the waste gas that is breathed out and builds up in the confined area of the air space around the mouth.

A snow rescue dog is able to detect the smell of an avalanche victim, to alert its trainer by barking, and then help to dig the victim out of the snow.

STILL BREATHING UNDER THE SNOW

In January 2005, 24-year-old Norwegian Martin Gulsrud was skiing in France when he was caught in an avalanche. He describes the experience: "I saw the snow raise up like a wave. ... I understood that I was going to be buried seconds later, so I started to expand my lungs so I could have space to breathe under the snow pack. At the same moment I crashed into this wall of moving snow, I was covered and the pressure started to increase. ... The pressure was enormous; I couldn't move a finger." He was found 20 minutes later under 8 feet (2.4 meters) of snow by rescuers. His chance of survival was improved because he wore safety equipment (a series of tubes with a mouthpiece strapped to his body) specially designed to divert carbon dioxide away from the mouth area.

MATERIAL STRENGTHS

Materials are more or less likely to slide depending on how strong they are. Stronger materials such as clay are stickier than weaker materials such as dry sand. Stronger materials have more friction between the particles than weaker materials. For example, saturated materials, such as very wet soil, are weak because the water forces the particles in the soil apart, reducing friction.

Equipment

- Equal amounts of fine soil, small pebbles, small rocks
- Water

- A deep plastic tray
- A cup
- A protractor

Instructions

1. Place the tray on a flat surface and pour the soil into one end of it.

2. Very slowly, raise the end of the tray containing the soil.

3. Stop lifting as soon as the soil begins to slide and hold the tray in position while someone else measures and records the angle of the tray with a protractor.

4. Repeat the steps for each of the other materials.

5. Repeat the above experiment, but before lifting, gently pour a cup of water over each material to thoroughly wet it.

6. Compare the results.

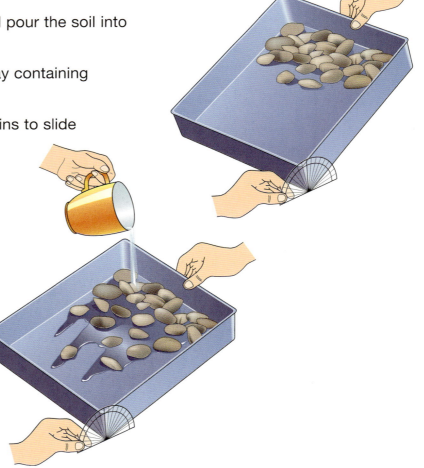

Alpine Of or relating to the Alps.

altitude A measure of height above Earth's surface or sea level.

canyon A narrow rift in the land with high, steep sides.

compact Closely and firmly packed together.

concave Hollow and curved, like the inside of a circle or sphere.

continental shelf The relatively shallow area of seabed that borders most continents, ending in a slope that descends steeply to deeper water.

crater A bowl-shaped hollow in the ground caused by an explosion, an impact, or an underground collapse.

creep The very slow, almost continuous movement of soil or other material down a shallow slope.

crust The solid outer layer of Earth.

debris Rubble, broken objects, and other damaged material.

debris flow A flowing mixture of water-saturated debris that moves downward under the force of gravity.

drought A long period of unusually dry weather.

earthquake A shaking of the ground caused by the sudden movement of underground rock.

ecosystem A system made up of a group of living organisms and their physical environment.

epicenter The point on Earth's surface directly above the center (focus) of an earthquake.

erosion A natural process by which rock and soil are broken loose from Earth's surface at one location and moved to another location.

eruption The pouring out of gases, ash, lava, and rocks from a volcano.

fault In geology, a break in Earth's crust.

fungus (plural: **fungi**) Organisms that obtain food by absorbing nutrients from other living organisms or from parts of formerly living things.

geologist A scientist who studies how the planet Earth formed and how it changes.

gradient The rate at which a slope goes upward or downward.

gravity The effect of a force of attraction that acts between objects because of their mass—that is, the amount of matter the objects have.

hurricane A tropical storm over the North Atlantic Ocean, the Caribbean Sea, the Gulf of Mexico, or the Northeast Pacific Ocean. Hurricanes are also known as *tropical cyclones* and *typhoons*.

lahar A volcanic mudflow, made up of water and ash.

lateral spread The sideways movement of material on gentle slopes or flat land following liquefaction.

liquefaction When soft, wet ground behaves like a liquid, such as during an earthquake.

loess A kind of silt that forms a topsoil in some parts of the world. It is made up of particles finer than sand but coarser than clay that have been brought by the wind to their present positions.

meltwater Melted snow and ice.

monsoon A wind that reverses itself seasonally, especially the one that blows across the Indian Ocean and surrounding land areas.

nutrient A nourishing substance.

precipitation Moisture that falls from clouds, such as rain, snow, or hail.

pumice Pieces of solidified, frothy magma that are full of air spaces. The spaces make pumice very light.

satellite An object that continuously orbits Earth or some other body in space. People use artificial satellites for such tasks as collecting data.

saturate To soak thoroughly.

saturation When the ground has absorbed all the water (or liquid) it can.

scarp A steep slope.

sediment Particles of earth, rock, or other matter carried along or deposited by water.

sewage Water that contains waste matter produced by human beings. Sewage comes from the sinks and toilets of homes, restaurants, office buildings, and factories. It may include harmful chemicals and disease-producing bacteria.

silt Fine-grained or muddy material made up of tiny particles of rock that settle at the bottom of rivers or other bodies of water.

slump The downward movement of blocks of material on a curved surface.

slurry A semifluid substance.

sonar A system that uses sound energy to locate objects; measure their distance, direction, and speed; and even produce pictures of them.

species A group of organisms that share certain permanent characteristics and can interbreed.

submersible An undersea vessel used for oceanographic research and exploration.

surface runoff Rainwater that runs across the land rather than soaking in.

terrain An area of land and its natural features.

topple The sudden collapse of material down a vertical rock face, such as a cliff.

tsunami A series of powerful ocean waves produced by an earthquake, landslide, volcanic eruption, or asteroid impact.

uplift The slow upward movement of parts of the seafloor or land.

volcano An opening in the crust through which ash, gases, and molten rock (lava) from deep underground erupt onto Earth's surface.

wildfire A fire that spreads quickly through forests and other vegetation.

BOOKS

Avalanche! The Deadly Slide, by Jane Duden, Perfection Learning, 2000.

Avalanche and Landslide Alert! by Vanessa Walker, Crabtree Publishing Company, 2004.

Erosion, by Virginia Castleman, Perfection Learning, 2005.

Landslides and Avalanches, by Terry Jennings, Belitha Press, 2003.

Landslides, Slumps, and Creep, by Peter Goodwin, Children's Press, 2002.

Volcanoes, by Anna Claybourne, Kingfisher, 2007.

WEB SITES

http://landslides.usgs.gov/learning/ls101.php

http://volcanoes.usgs.gov/Hazards/What/Landslides/landslides.html

http://www.disastercenter.com/guide/landslide.html

http://www.pbs.org/wgbh/nova/avalanche/

http://www.slf.ch/info/Caution_Avalanches.pdf

www.avalanche.org/

INDEX